HEADStart

counting

"Hello, I'm Rachel and this is Suni."

"Come and help us find out about counting."

Linda Fisher

Illustrated by Simone Abel

BROCKHAMPTON PRESS
LONDON

NOTES FOR PARENTS

When children first start counting, they tend to point indiscriminately at the objects they want to count while chanting numbers. Try to help your child understand that each object is counted only once, is given only one number name and that the last number reached is the total. Your child will understand this better if you put each object to one side as it is counted, or cross each one off or cover it up if you are counting items in the book.

Being able to recite numbers in the correct order is a necessary skill, but your child also needs to begin to understand how numbers actually work and to recognise the patterns in them. He or she needs to know how many '3' actually is, that '4' is more than '2' and that if 5 children are each to have a drink, then 5 mugs will be needed.

The activities in this book are designed to give practice and understanding in these number skills and to increase your child's number vocabulary by using terms such as 'more', 'long', 'short' and 'the same as'. You can also help your child with numbers in your everyday activities together. Ask questions such as, 'How many steps do we have to take to reach the front gate?' and, 'Will all your bricks fit into this box?'.

At this stage your child is not ready to do 'sums', but is building up a good store of basic understanding ready for when more formal education begins.

Pages 4–5 Balloons
Here your child will be counting and matching two balloons to each child on the page. Use this activity to introduce the phrase 'the same as'.

Pages 6–7 Whose shoes?
Let your child try on all the family's shoes. Talk about sizes such as too big, too small and just right.

Pages 8–9 Ladders
This page takes another look at sizes – this time 'long' and 'short'. Help your child draw in the ladders by first deciding how long each one should be.

Pages 10–11 Picnics
This activity lets your child match on a one to one basis. Help your child count the number of picnickers and the number of plates and drinks in each picnic to find out which ones match. Count the number of people in your family together before your child draws in the family picnic.

Pages 12–13 Glove hunt
Here your child needs to find the pairs of gloves first before trying them on by looking at the number of fingers and thumbs on each glove. The left-hand page has the left-hand glove on it, the right-hand page has the right-hand one on it.

Pages 14–15 Bee game
This game gives more counting practice. Finding out how many are left forms the beginning of understanding subtraction.

Pages 16–17 Where do we live?
Sorting into sets or families is a basic stepping stone in number work. Talk to your child about which animals live where.

Pages 18–19 Let's go for a walk
Your child may not be ready to understand about distances, but help your child to describe his or her journey using phrases such as, 'Down the short road, over the bridge and round the corner.'

Pages 20–21 Milkman game
This game encourages your child to match objects – the milk bottles – to numerals.

Pages 22–23 All or some
Your child will probably understand the term 'all', but you may need to help him or her to understand that 'some' is variable and not a specific part of a whole.

Page 24 Finish the patterns
Numbers make repeating patterns, and this activity helps your child recognise patterns and predict what comes next, an important stage in number work.

3

Balloons

Can you draw the balloon strings so that all the children have 2 balloons each?
The first one has been done for you.

Can you draw another balloon for each child?

Join the bunches of balloons which are the same.
Finish colouring the balloons.

Whose shoes?

Look at all these feet.
Can you draw a line from each pair of feet to the shoes which will fit them?

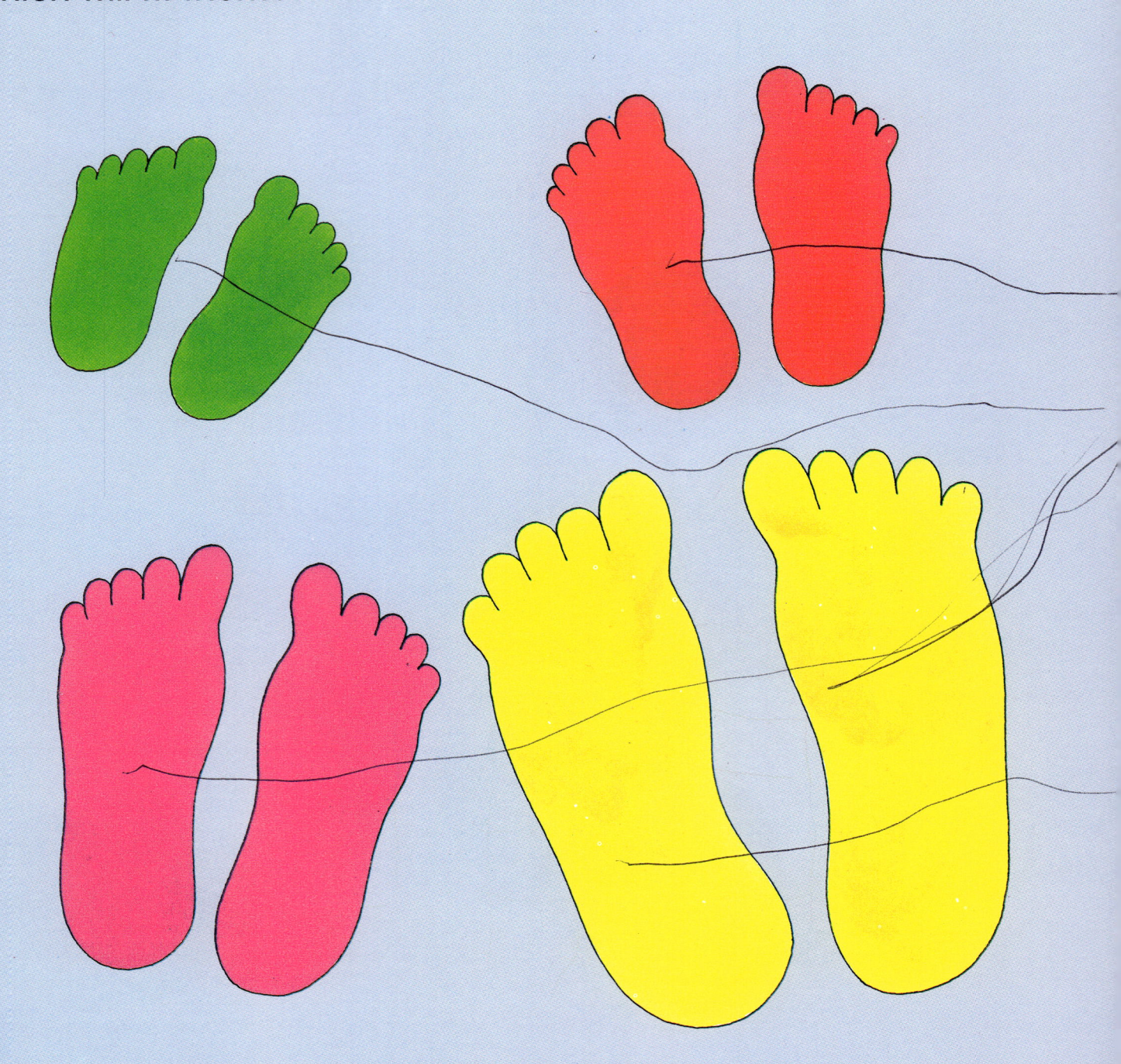

6

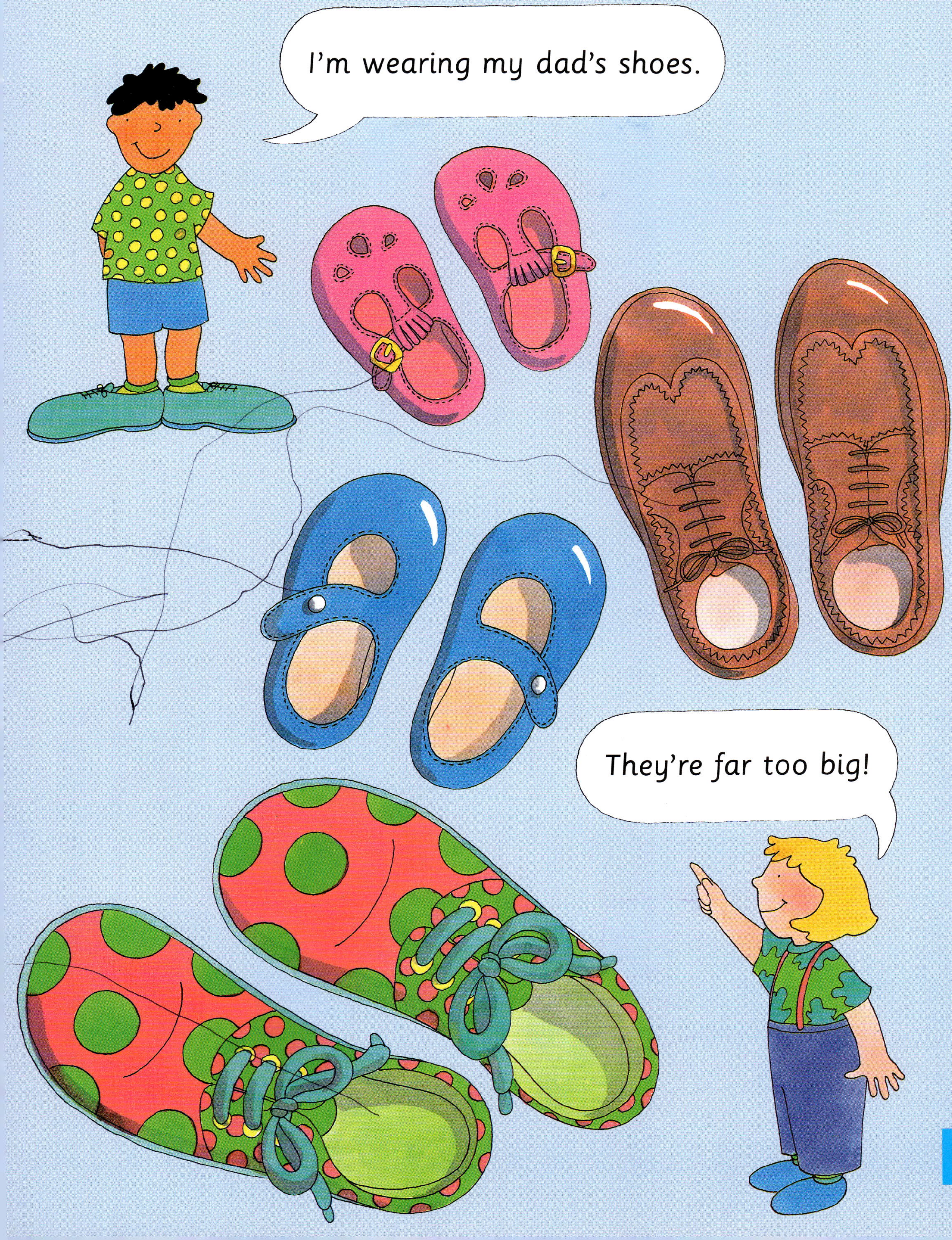
I'm wearing my dad's shoes.
They're far too big!

Ladders

Here is a long ladder.   Here is a short ladder.

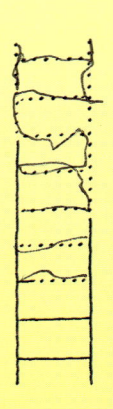

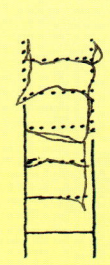

Finish drawing each ladder.

Practise drawing some more ladders here.

Help these busy people by drawing in ladders for them.

Picnics

The teddy bears are having a picnic. Can you draw the paths so that the bears can reach their picnics? Make sure they each have a plate, a drink and a cake.

Draw a picnic here for your family.
Give each person a plate, a cup and a cake.

Glove hunt

We're trying on gloves.

Try on these gloves.
Can you find a pair which fit you?

13

Catch the bees

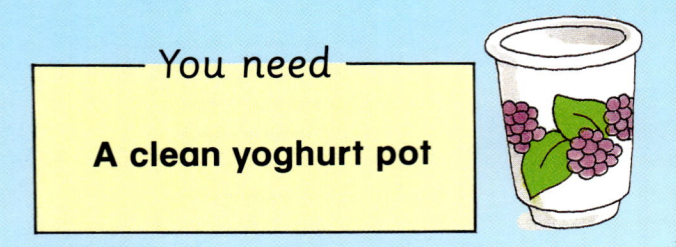

You need

A clean yoghurt pot

Can you use your pot to catch:

1
2
3

If you can catch 4 all at once, tick (✔) here:

Be careful you don't get stung!

Now play this game with a friend. Both of you catch some bees. How many are left?

15

Where do we live?

Draw a line to join each animal to its home.

17

Let's go for a walk

We're taking our fingers for a walk.

Walk your fingers from the [Bakery/Dress Shop] to the [phone box].
Was it a long way?

Now walk from the [Bus Stop] to the [house].
Was this longer or shorter than your first walk?

Can you find a very long walk and a very short walk?

Remember to cross the roads safely in the correct place .

Milk round

Help the milkman deliver the milk.
Can you draw the right number of bottles of milk on each doorstep? Each door has a note to tell you how many bottles are needed.

I've done the first one.

3 today, please.

4 today, please.

5 today, please.

All or some

All of Suni's friends have toys.

Some of these trees have leaves.

Can you draw spots on all of these dogs?

Draw eggs in some of these egg cups.

Washing day

We've pegged the clothes on the washing lines to make some patterns.

Draw what comes next in each pattern.